Genre Nonfiction

Essential Question
How do machines make our lives easier?

Machines

by JANET PALAZZO-CRAIG

Chapter 1
Machines at Work

The world is a busy place! Look around you. People are riding on skateboards and driving cars. They are climbing stairs, lifting packages, and digging holes. They are all doing some kind of work.

People have made special tools. The tools help you do work. These tools are called machines.

There are **simple machines** and **compound machines**. Simple machines are the most basic. They have few or no moving parts. Compound machines are two or more simple machines put together.

A skateboard is a compound machine made up of simple machines.

Each of these students is doing work.

Work It!

You do work. That means moving something or changing its motion. You move dirty clothes into a laundry basket when you clean your room. When you take out the garbage, you move bags to a trash can.

To do work, you must use some effort, or force. The effort for simple machines comes from your muscles when you push or pull. The effort for compound machines may also come from your muscles. It can come from other sources, too. It might come from wind or electricity.

A ramp can help you do work.

How Do Machines Help?

Machines help you do work. A simple machine does not really cut down on the work it takes to do a job. It just lets you use less force. There is a trade-off, though. When you use less force, you have to apply that force over a longer distance.

Think about lifting your bike up and into a truck. That is hard work. Now think about placing a ramp from the ground to the truck. You don't have to lift the bike up. You just push or pull it up the ramp.

The ramp is a simple machine. It is called an **inclined plane**. The push or pull you use to move the bike is the effort force. You use less **effort force** when you use the ramp. You must move the bike farther than you would have if you lifted it up, though. This means there is a trade-off between the amount of force used and the distance moved to do the work.

You do the same amount of work in each case. The work seems easier using the ramp. This is because you use less effort.

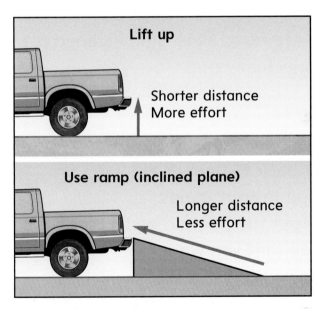

Using a ramp (inclined plane) to move something takes less effort force than lifting it even though the distance you move it is longer.

Chapter 2
The Six Simple Machines

There are six kinds of simple machines. The six machines are the inclined plane, the **lever**, the **wheel and axle**, the **pulley**, the **screw**, and the **wedge**. All machines are made of one or more of these simple machines.

The Inclined Plane

The inclined plane is the simplest. It has no moving parts. It is a flat, slanting surface. An inclined plane makes it easier to move things up or down. A playground slide is an inclined plane, and so is a wheelchair ramp.

The Wedge

The wedge is two inclined planes put back to back. You use a wedge to split things apart. You put the pointed end into a space that you want to open. Then you push on the wide end. The wedge changes the downward force. It becomes a sideways force. The blade of an ax is a wedge.

Wedge

The Screw

The screw is really an inclined plane. It has a slanted surface. It is wrapped around a center pole. This slanted surface is an inclined plane. It lets the screw move forward when you turn it.

The Lever Gives You a Lift!

The lever helps you lift or open things. This simple machine is a bar that rests on a fixed point called a fulcrum. The fulcrum lets the bar turn. When you use a lever, you apply an effort force. The object you are moving is called a load. A bottle opener is a lever. A shovel is a lever, too.

Screw

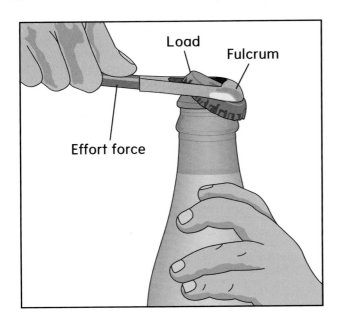

Load
Fulcrum
Effort force

Lever

Pulling with the Pulley

The pulley is a wheel with a rope or chain around it. It helps you lift things. It does this by changing the direction of your force. It works because pulling down seems easier than lifting up. To use a pulley, you tie a heavy load to one end of the rope. You pull down on the other end. The heavy load lifts up.

A pulley makes it easier to lift heavy objects.

Rolling with the Wheel and Axle

You can think of the wheel and axle as a lever that turns in a circle. A wheel is connected to a rod called an axle. The wheel turns with the axle. This simple machine makes it easier to move or turn things.

The wheel is larger than the axle. As it turns, the wheel takes less effort force to move than the axle does. It moves a longer distance, though. The axle takes more force to move, but it moves a shorter distance.

A doorknob is a wheel. It connects to a rod. This is an axle. You could open the door by turning the knob or the rod. The knob takes less effort force to turn than the rod does. This makes opening the door easier. You must turn the knob farther than the rod, though.

8

Jose A. Bernat Bacete/Getty Images

Chapter 3
Simple Machines Form Compound Machines

When two or more simple machines work together, they form a compound machine. Bicycles, hand drills, and escalators are some compound machines you may use.

The Can Opener Can Do It!

Did you know that a can opener is a compound machine? It is made of these simple machines:

- Lever (The hinged handle)

- Wheel and axle (The turning knob)

- Wedge (The sharp blade that cuts the metal can)

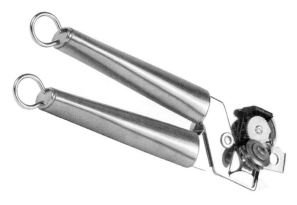

A can opener is a compound machine made up of simple machines that work together.

The Escalator: Step Right Up!

Have you ever taken a ride up the escalator? If so, you have used another compound machine. An escalator takes the work out of walking up or down stairs. An escalator gets power from an electric motor.

These simple machines make up an escalator:

- Inclined plane

- Wheels and axles

- Pulleys

How do these simple machines work together in the escalator? First, think about the slant of the steps. The steps form an inclined plane.

Simple machines work together in this compound machine.

Next, look inside the escalator for the wheels and axles, and pulleys. You'll see something that looks like the chain on a bicycle. A group of wheels is stretched around two large gears. They are below the escalator's steps. Gears are wheels with teeth on them. The teeth are small wedges. They can move other objects as they turn.

The escalator's gears turn and move the wheels in a circle under the steps. This moves the steps up or down an inclined plane. The electric motor is attached to one of the large gears with a pulley. As a pulley turns the large gear, the wheels under the steps move. The wheels and gears work like a wheel and axle. Pulleys also move the handrail of the escalator.

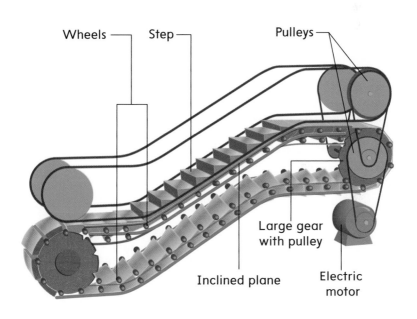

Wheels — Step — Pulleys —

Large gear with pulley

Inclined plane

Electric motor

Respond to
Reading

Summarize

Use details from the text to summarize *Machines*. The graphic organizer may help you.

Cause → Effect	
	→
	→
	→
	→

Text Evidence

1. What are simple machines? What are compound machines?

2. Read the book again. Read with a partner. Talk about how simple machines help you do work. This is the effect. Fill in a cause-and-effect chart with your partner. Show the causes. CAUSE AND EFFECT

3. Use what you know of antonyms. Figure out the antonym of *same* on page 5. ANTONYMS

4. Write a letter to a movie theater owner. Ask the owner to put in a wheelchair ramp. Explain the effects of putting in the ramp. Tell why it is important. Use words to persuade, such as should and must. WRITE ABOUT READING

Compare Text

Read about how Pedro and Mia used simple machines to help build a playground.

Made with Machines

Pedro and Mia were excited. They were helping to build a playground.

"Can you kids bring the boards up here?" a man called. He was standing high up on a platform.

"We can do it!" Pedro answered

Pedro and Mia tried to put heavy boards into a wheelbarrow. The wheel made it easy to move the boards.

Now they had another problem. How could they get the boards up to the platform?

The man said, "Put the boards in the bucket." Pedro and Mia loaded the bucket.

The man used a pulley. He raised the bucket full of boards to the platform.

Next, Pedro and Mia used the wheelbarrow to move cans of paint. Pedro was worried. "The wheelbarrow won't go up stairs." he said. "I have an idea." he added.

Pedro put a plank over the stairs. They rolled the wheelbarrow up the ramp. Soon Pedro and Mia were helping to paint the playground.

Then the playground was finished. Pedro and Mia were proud. Pedro said, "We couldn't have done it without our simple machines."

Make Connections
How did the children use machines?
TEXT TO TEXT

Glossary

compound machine *(KOM-pownd muh-SHEEN)*
a combination of two or more simple
machines *(page 2)*

effort force *(EF-uhrt FORS)* a push or pull you apply
to a simple machine to move something *(page 5)*

inclined plane *(in-KLINED PLAYN)* a simple
machine formed by a flat, slanting surface *(page 5)*

lever *(LEV-uhr)* a simple machine made of a bar
and a fixed point, called a fulcrum *(page 6)*

pulley *(PUL-ee)* a simple machine made up of a
rope or chain wrapped around a wheel *(page 6)*

screw *(SKREW)* a simple machine that is an
inclined plane wrapped into a spiral *(page 6)*

simple machine *(SIM-puhl muh-SHEEN)* a machine
with few or no moving parts, making it easier to
do work *(page 2)*

wedge *(WEJ)* a simple machine that uses force to
split objects apart *(page 6)*

wheel and axle *(WEEL and AK-suhl)* a simple
machine made of a rod called an axle attached
to the center of a wheel *(page 6)*

Index